For Ry
My 'fun guy'
xxx

Copyright © 2023 by Nicola Robson. All rights reserved.
No part of this book may be reproduced in any form or by any electronic or mechanical means, including information storage and retrieval systems, without written permission from the author, except for the use of brief quotations in a book review.

ISBN 9798387245497

As the world is getting faster
it's important that you know,
to stop and look and listen,
to move steady and go slow.

We have everything we need
sitting right beneath our feet,
if we care for mother nature
then the circle stays complete.

Always enjoy & respect nature.

Never eat any wild food without multiple sources of positive identification. Some wild plants and mushrooms are poisonous and should not be touched or consumed.

Magnificent Mushrooms

In an enchanting kingdom beneath the ground, there lays a mighty organism that has the power to heal the world.

Let's go on a journey into the mysterious world of mushrooms!

As you step inside the forest
and take a look around,
with every tiny footstep
there are secrets to be found.

With branches reaching to the sky,
the mighty trees are tall,
but there's something in the forest
more important than them all.

You'll find them under crunchy leaves
and hiding in the dark,
they're growing over fallen trunks,
decomposing rotten bark.

The humble little mushroom,
more essential than you know,
they keep the forest nourished
and they help the trees to grow.

If you keep your eyes wide open,
you'll spot the mushroom signs,
from chanterelle to puffball,
twenty thousand different kinds

Some have very funny names
like chicken of the woods!
From jelly ear to penny bun
and some have spotty hoods.

They are experts at recycling
with a very special role,
giving life back to the soil
is a mushrooms only goal.

Fungi are surprising,
they work without a sound,
Communicating with the trees
through threads sent underground.

Threads that grow for miles,
like a secret little web,
a network of mycelium,
under ever single step.

Mycelium holds the key to life,
it grows from tiny spores,
released from all the mushrooms
that line the forest floor.

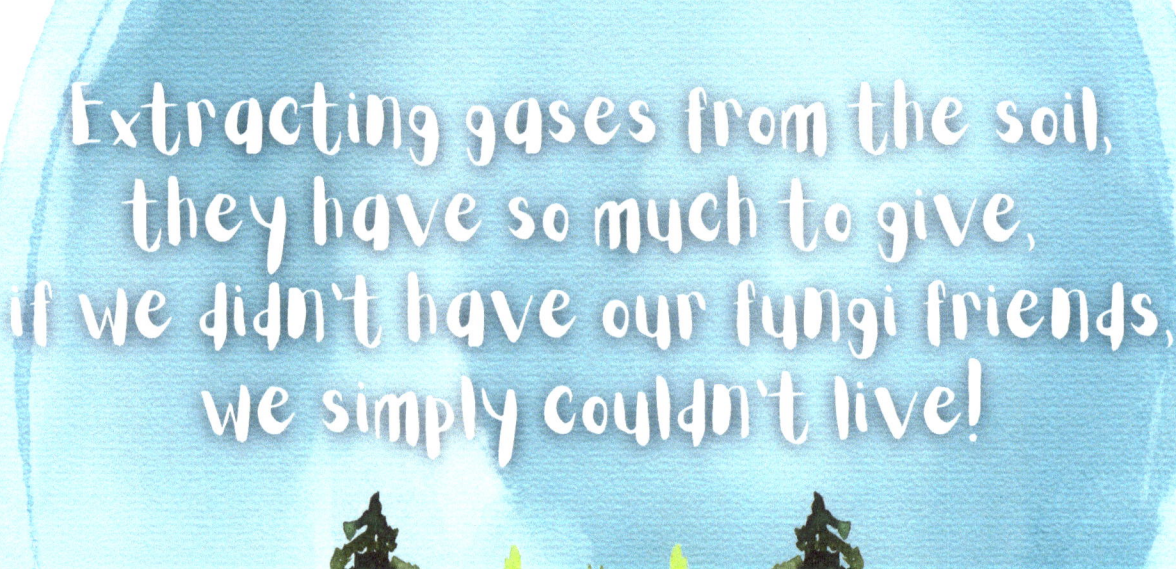
Extracting gases from the soil,
they have so much to give,
if we didn't have our fungi friends,
we simply couldn't live!

They feed entire forests,
from great oaks to tiny flowers,
enriching our amazing earth
is a mushroom's greatest power.

If mushrooms weren't around to decompose the leaves, mounds of dead plant matter would swallow up the trees.

This simply wouldn't work for us,
as if we had no trees,
the eco system couldn't last,
no living thing could breath.

They really are quite special,
more than they might seem,
mushrooms have the power,
to keep the whole world clean.

They can break down bits of plastic,
and clean up oil spills,
they can freshen up the waters
with extraordinary skill.

There are even types of mushrooms
that glow up late at night,
bioluminescent wonders,
a most enchanting sight.

These natural little night lights come in many different forms, attracting night time insects that help to spread their spores.

The oldest and the youngest
of all the life we know,
the first ones here on planet Earth,
and they'll be the last to go.

In a world of great invention,
the answer is quite clear,
mushrooms are the greatest gift
that ever did appear.

So now it's down to you,
to help them do their thing,
take care of all of mushrooms,
watch the magic that they bring.

New trees will line the forests,
fresh air will fill our lungs,
the world will be in balance,
a future brighter than the sun.

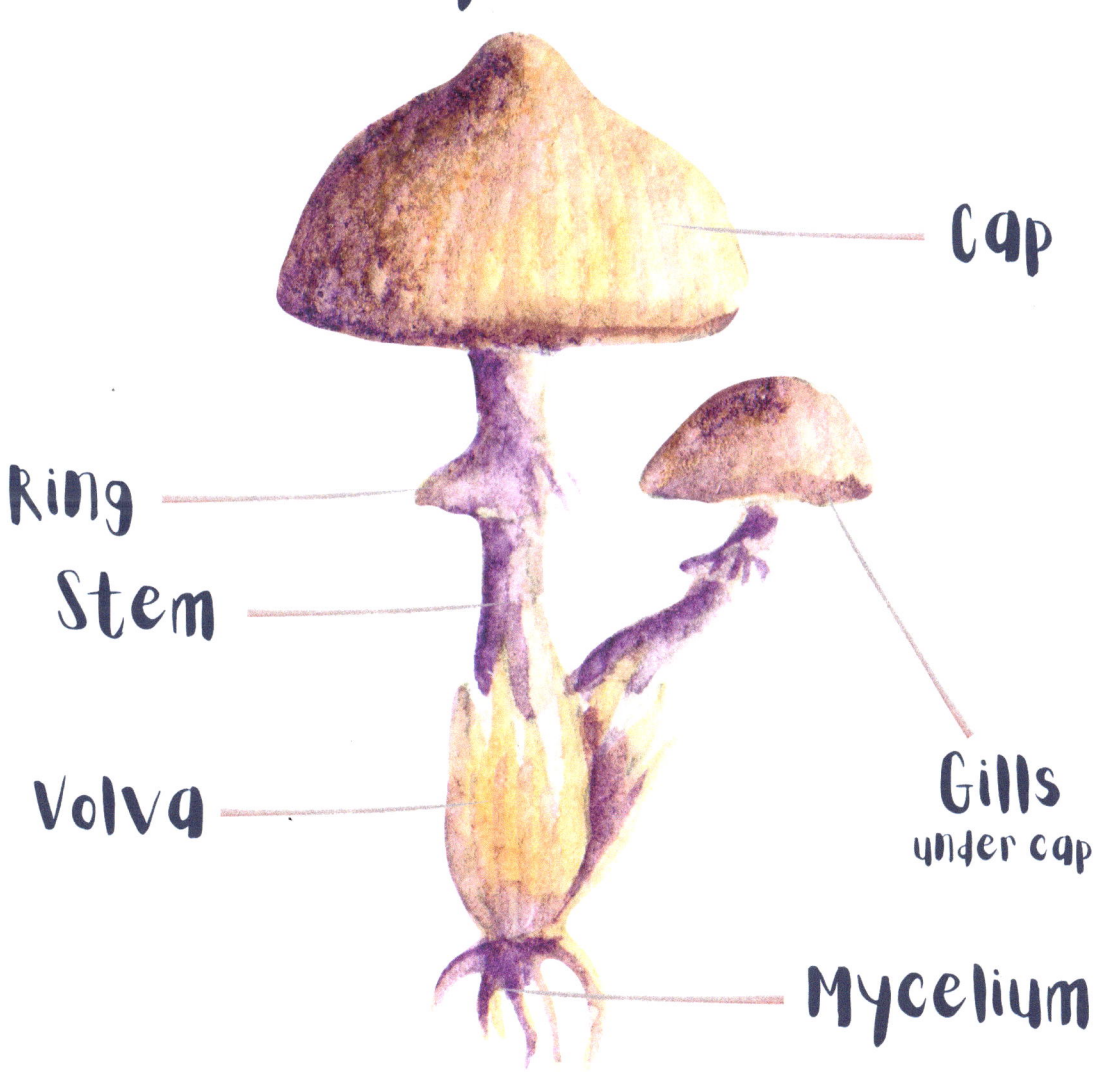

Did you know...?

'Mushrooms are not a vegetable!'

Mushrooms are part of the Fungi family. They are not a vegetable or a plant, they are from a kingdom of their own.

'There is an edible mushroom that tastes just like fried chicken!'

Chicken of the woods is its name and it can be found across Europe and North America.

'Mushrooms help trees talk to each other!'

Through a network of threads underground called mycelium, mushrooms connect all the trees in the forest and help them comunicate.

'Mushrooms can break down plastic!'

Plastic will take about 400 years to decompose on its own but some mushrooms are able to break it down in just a few months.

'The largest living organism on Earth is a fungus!'

The giant Armillaria Ostoyae was discovered in the Blue Mountains of Oregon in 1998 and is the size of 1,665 football fields!

'There's a mushroom that eats itself!'

Once picked, the Shaggy Incap's black goo drips to the ground to release its spores then uses auto digestion to devour itself.

Ways you can
take care of nature...

Plant a tree.

Buy pre-loved toys where you can.

Before you throw something away think about how you can re-use it.

Recycle what you can't re-use.

Don't use only one side of your paper when you draw pictures.

Take your litter home with you.

You can make a difference!

www.blossomandstardust.com
@blossomandstardust

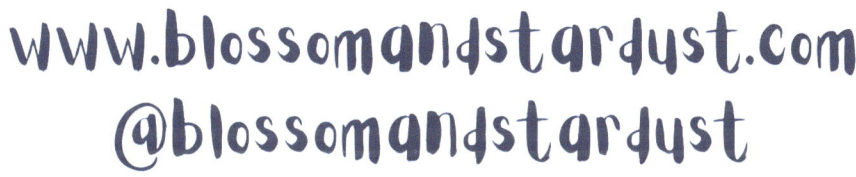

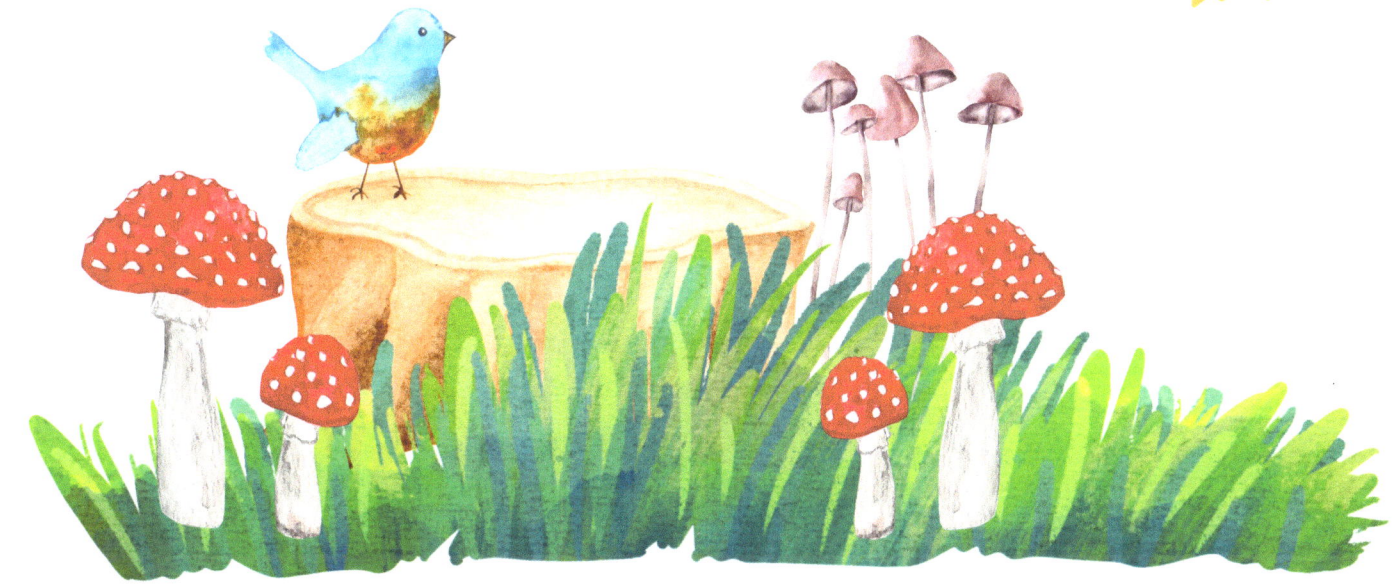

Check out my other books...

www.ingramcontent.com/pod-product-compliance
Lightning Source LLC
Chambersburg PA
CBHW051937210526
45473CB00006B/2276